WAYNE'S WORLD *of* PHYSICS

WRITTEN BY
CARREL W. UPTERGROVE

To order additional copies of this book, contact:
Xlibris
1-888-795-4274
www.Xlibris.com
Orders@Xlibris.com

The following articles in *Wayne's World of Physics* were copyrighted by Carrel W. Uptergrove.

Article 1. Why Holland Needs Dikes, copyright ©, TXu 922-717, on Oct. 29, 1999

Article 2. A Rebuttal of Big Bang Theory, copyright ©, TXu 918-977, on September 7, 1999

Article 7. Hubble Red Shift, copyright ©, TXu 597-391, on August 26, 1993

Article 8. Infinitely Aged Universe ©, TXu 594 617, on September 10, 1993

Introduction

I am publishing this book to share some of the things I've learned that you might not be aware of. For a long time, I have said that there is an infinite knowledge in the universe that man does not know yet. It would be a mistake for me to leave this world and not pass on the information I've discovered during my study and research results.

Long after I wrote, "Why Holland Needs Dikes", I realized that because the moon goes around the world from west to east, it is a large contributor to the force of the large current that flows through the Indian Ocean and around Antarctic.

In the, "Rebuttal of the Big Bang", I did not address entropy because I do not believe it exists in the dynamic exchange between electromagnetism and gravity creating the stars. These two forces are much too powerful to be affected by entropy.

"Fresh Water" was written to try to get a dialog going with entities building desalination plants. I have designed a way to get air back into the air chambers without the use of fossil fuels.

I wish to thank my wife, Mary Ellen McCullock Uptergrove, for her skill in word processing and for the many hours she spent bringing my articles into book format. She is the light of my life and a joy in my heart.

Carrel Wayne Uptergrove is the name I came into this world with. Carrel is the Danish version of Charles and is pronounced the same as carol, the joyful singing we do at Christmas. I was always addressed with my first and middle name at home and at school until about the middle of my time in elementary school, then I became Wayne. Carrel W. remains for all my legal requirements and signatures.

I hope you enjoy the articles, especially "Why Holland Needs Dikes".

Carrel W. Uptergrove

Contents

Article 1. WHY HOLLAND NEEDS DIKES
By Carrel W. Uptergrove

In 1994, a radar satellite (ERS-1) (Aviation Week and Space Technology: October 24, 1994) documented the topography of ocean levels worldwide. The startling revelation was that ocean levels vary by an unbelievable 627 feet.

The lowest ocean levels are in the Indian Ocean surrounding India, while the North Atlantic, North Pacific and Arctic Oceans are much higher in elevation than the Indian Ocean. Several other areas worldwide are elevated in comparison to the Indian Ocean as well.

When one considers that the combined gravitational tug of the sun and moon on the earth's oceans create tides of only a few feet and that the highest tides on earth are created in the Bay of Fundy on the order of 30 feet, then it becomes obvious that gravitational anomalies are not the cause of ocean elevation differentials of this magnitude.

Why does Holland need dikes to hold the elevated North Atlantic Ocean at bay? There are several reasons and we will address each reason in turn. To oversimplify, we could say that Holland needs dikes because:

1. The sun is plasma and has a magnetic field.

2. The earth has a solid iron core in its center and obeying the laws of electromagnetism as explained by James Clerk Maxwell, an electric current is induced into the iron core as the earth cuts the magnetic lines of force in its orbit of the sun.

3. The induced current in the iron core drives the earth to rotate on its axis exactly as an induction motor. The iron core induction motor drives (causes to rotate) the molten magma, the earth's crust, the oceans and the atmosphere.

4. The earth has a unique land mass configuration forcing the ocean currents flow to the northern oceans.

It takes a powerful force to hold the northern oceans of the world 627 feet above that of the Indian Ocean. That force is the magnetic field generated by the Sun and the interaction of that magnetic field with the earth's iron core.

The earth's iron core is on the order of 3,000 miles in diameter. As the iron core moves around the sun, it cuts the sun's magnetic lines of force and an electric current is set in motion within the core. Then the effect is that the core becomes an induction motor and aids the earth's rotation on its axis.

If the earth did not already rotate on its axis, the induced electric current in the core would set the earth to rotating on its axis.

The iron core is a nearly perfect sphere with a small crack near its surface about 1,000 miles long and running north to south. The crack was discovered about 30 years ago at about the longitude of St. Johns, Newfoundland, by using seismic sounding.

It is because of this crack that we can detect the movement of the iron core with respect to the surface of the earth. During the 30 years that we have monitored the crack, it has moved eastward until it is just south of Greenland. This is proof that the earth's iron core acts as an induction motor and spins the earth's iron core faster than the rest of the earth.

The earth's surface spin is coupled to that of the iron core by a viscous magma surrounding the iron core. The oceans are in turn coupled to the spin of the earth's crust and the atmosphere is coupled by the spin of the oceans.

If you wish to feel the atmosphere being dragged around the earth, you can do so by going to sea and at the equator, facing east, you will feel a 15 to 30 mph wind in your face. You feel this wind because at the equator you are traveling eastward at approximately 1,000 mph.

If the earth had not been rotating on its axis 5 billion years and dragging its atmosphere with it imparting an eastward rotation of the atmosphere with it, you would feel a wind of 1,000 mph. Instead what you feel is a 15 to 30 mph wind as the friction between land, sea, air, and your own body drags the atmosphere eastward with the driving force of the earth's iron core induction motor.

THE OCEANS ARE ALSO BEING DRAGGED EASTWARD BY THE SEA FLOOR AND THE CONTINENTAL LAND MASSES.

It is precisely this dragging of water and air that cause the oceans to pile up in the equatorial regions of the Pacific near Australia, New Guinea, Solomon and other islands.

This pile up of water at the equator would be released equally to the north and south except for the fact that the Solomon and New Guineas Islands are physically situated so that a disproportional amount of pressure is released northward. A vast amount of water is forced to flow clockwise around the North Pacific elevating the Pacific and tending to elevate the Arctic Ocean as well.

As we shall see in further discussion, however, a much greater water pressure already elevates the Arctic Ocean from the North Atlantic.

We must first discuss some of the other ocean currents because we will see that, in part, some of the energy and force necessary to elevate the entire North Atlantic and Arctic Oceans comes from these currents.

The South Pacific current describes a counterclockwise motion and compared to the North Pacific is a smaller current. However, at the southern portion of its circuit this current joins and becomes part of the Antarctic current. This current around Antarctica is so powerful that it stirs up nutrients from the sea floor that supports a prolific sea life. It is said that this current supports 90% of all life on earth because of the nutrients stirred up from the sea floor.

Examining the topography of the sea floor southeast of Argentina evidence is seen of a vast alluvial plain deposited from land which preciously connected South America (Drake Passage) to the Antarctic and is now deposited on the sea floor by this powerful current.

The velocity of the Antarctic current is sufficient to force the current over the shallow areas southeast of Africa (Kerguelan Island is in this area). The high elevations of sea levels in this region are caused by this high current in the same way that water is elevated over a boulder in a swift flowing stream.

As the current leaves this elevated area, it speeds up on its journey past the southern end of Australia and a Venturi effect is created drawing water from the Indian Ocean.

It is this Venturi effect and the consequences of drawing water from the Indian Ocean that causes the waters south of India to be the lowest seas in the world.

It is now time to discuss the Atlantic current. Let's begin by going back to Drake Passage and following the Antarctic current eastward. As the current nears the southern tip of Africa, it begins to feel the high pressure east of Africa due to the shallow seas and some of the current is deflected northward toward the equatorial region of Western Africa (Ivory Coast).

As the current hits the Ivory Coast, a mound of water builds up and the current is deflected westward along the equator. The prevailing winds now carry the current toward the South American coast.

The current does not feel the continental land mass until it has passed Cabo de Sao Roque in Brazil which lies five degrees south of the equator and as a consequence most of the current is deflected north into the North Atlantic and becomes known as the Gulf Stream.

It is this large and powerful current that now flows to Europe and the Arctic Ocean elevating these ocean levels 627 feet higher than the waters in the Indian Ocean.

That is why Holland needs dikes.

Article 2. A REBUTTAL OF THE BIG BANG THEORY
By Carrel W. Uptergrove

"I want to know how God created this world. I am not interested in this or that phenomenon, in the spectrum of this or that element. I want to know His thoughts; - the rest are details." Albert Einstein.

PREFACE

This is written to help the ordinary reader understand the concept of the big bang theory and the laws of physics used to expound or refute the argument. The reader should have some basic concepts about physics and cosmology.

I would like to say that if we think about all of past history, many of the concepts and beliefs as proposed by the most learned scholars of each previous period of time have been proven wrong.

I propose that the scholars are more wrong in their concept of the cosmos now than in all of man's history. I would prefer to believe that the stars I see in the heavens are the campfires of nomadic tribes wandering the heavens than to believe the big bang theory.

LIGHT INTENSITY DECREASES WITH THE SQUARE OF THE DISTANCE

As a young man, Einstein made the observation that the sky was dark at night. Using his technique of though experiment he reasoned that the universe must be finite in size because if it was infinite then the sky would be brilliantly lighted by an infinite number of stars.

But this presented another problem because if it was finite in size then the stars would have a tendency to gravitationally collapse together into a small region of the universe.

Since he knew of no evidence that this was occurring he inserted a force (cosmological constant) in his equations, which was equal to the gravitational attraction of the universe and acted in the opposite direction of gravity so that the universe did not collapse upon itself. This force was called the cosmological constant.

As a young man I made the observation that the sky was dark at night. I also reasoned that a star would become invisible to me at some distance because light spreads out in all directions from the star and because of this the brightness becomes dimmer with distance. Just as a book I read at night by coal-oil light became dimmer the farther away from the light I moved.

Coal-oil light intensity is rather weak so one notices distance more readily than one would by reading using modern lighting. Indeed, in basic physics I learned that the brightness (intensity) decreased with the square of the distance from the light. This squaring of distance decreases the intensity dramatically. For example: 2x2=4, 4x4=16, and 16x16=256.

So light falling on the page of a book is decreased 256 times in intensity at 16 feet, in comparison to the light at 1 foot. Does it not seem reasonable to assume that at an infinite distance light intensity from a star would never reach earth.

Intensity would become infinitely small because infinity squared is a very large number. Indeed, when it is not squared, it is a very large number. If this tenet of physics is not correct, Einstein should have addressed and corrected the equation.

This intensity effect of the square of the distance equation can be made visually graphic if the reader will take a floor lamp with at least a 200-watt light bulb (clear glass is better) in it. Remove the lampshade and place the light in the approximate center of the room during darkness.

Turn on the lamp and notice how the entire room is illuminated. Next, place a book against the lamp and notice how one half of the room (universe) is illuminated and one half of the room (universe) is in the shade of the book. Half of the energy from the lamp falls on the book and half on the rest of the room (universe).

Any light that you see in the shade of the book is reflected light from the illuminated walls and objects in the room. The shade of the book would be totally dark if there was no reflected light from walls and objects in the room.

Now move the book slowly away from the lamp watching the shadow of the book rapidly decrease in size with increasing distance from the lamp. Finally, notice that as the book almost touches the wall, the shadow of the book has decreased dramatically and a much larger portion of the room (universe) is now directly illuminated.

The lamp except for the small shadow behind the book now directly illuminates the room (universe). The reader can readily see that the book has far less light from the lamp falling on it and that light now directly illuminates a much larger portion of the room (universe).

Now, in your imagination, continue to move the book to an infinite distance from the lamp and visualize that the book is an infinitely small object and receives no light.

So it is with the human eye. It receives no illumination from a star placed an infinite distance from it.

RED SHIFT OF LIGHT FROM DISTANCE
GALAXIES OBSERVED BY HUBBLE

Hubble, by using the powerful telescopes, which became available in his day, observed a red shift in the distant galaxies he could observe with these wonderful new telescopes. That is to say that the galaxies appeared red to the eyes as one observed them, much the same as the sun or moon take on a red appearance to the eye when low on the horizon at sunset.

The farther away the position of a galaxy was in any direction the redder it appeared. Hubble explained this phenomenon by hypothesizing that it was as if the galaxies were moving away from us, and a Doppler effect caused the red shift much as a train whistle changes pitch as it recedes from you. It was reported that Einstein was delighted.

Quasi-particles as described in quantum mechanics and the Compton effect (equation) will be presented herein as more valid causes for the red shift observed by Hubble.

THE BEGINNING OF TIME

Gamow, making it a holy trinity, made the suggestion, that thinking backward in time there must be a beginning to time, that the expansion must come from a big bang (explosion) from an object as small as a singularity – a singularity that was much too small to see.

As a young man, I concluded that the universe was infinite in size, extended in an infinite direction from me and must be infinitely aged. No other conclusion seemed rational.

Nevertheless, and disregarding my conclusions, I have spent a large time of my life trying to believe and understand what this trinity suggests and indeed, something the entire world seems to believe.

However, embedded deep in the fundamentals of physics and in my own thought experiments, I find valid reason to reject this theory.

Besides the decrease of light intensity with the square of the distance already mentioned some of these fundamental reasons are as follows:

THE CORNERSTONES OF THE BIG BANG THEORY
The big bang theory rests mainly on four cornerstones, which are:

1. The Doppler red shift observed by Hubble.

2. The microwave radiation background.

3. The distribution of the elements.

4. Entropy

If the first reason can be explained using well-accepted fundamentals of physics, other than a Doppler effect, then one cornerstone should be demolished. If the microwave radiation background can be explained with a reason other than residual primordial heat left over from the Big Bang; and if the distribution of the elements can be explained in an infinitely aged universe; then the big bang theory cannot stand; just as a building cannot stand with three cornerstones knocked away. Let us consider the three cornerstones in the order presented above.

THOUGHT EXPERIMENT

Let us do a thought experiment. First, we will create the equivalent of the human eye looking at the equivalent of a white star so that we can easily understand why a white star appears white if near to us like our sun and appears more red-shifted at great distances. Our eye will have three different types of receptors; several will receive red light.

With these three types of receptors the eye will perceive and see all the colors in the rainbow, just as color television uses these colors to achieve the same effect. Any wavelength other than the colors of the rainbow cannot be seen by a human eye or by our equivalent eye. We have now defined our equivalent eye, which interprets light very much as the human eye does.

Now we only need a white star much like our sun for the eye to see. The white star will emit only the three colors, blue, green and red. This is done to simplify the discussion and the results will be identical to a real white star, just as a television picture tube uses these three colors and achieves the same result as transmitting the entire rainbow.

Our white star emits only these three colors and no radiation is emitted outside of the visible spectrum. The reader should remember that from these three colors come all the colors of the rainbow, including black and white. Black is the absence of any color and white is the presence of all colors.

The eye perceives color as a definite number of energy quanta for that color and each color has a different number of energy quanta. Planck, who spent six years of his life discovering and defining what we now call Planck's constant, states that light comes to us in bundles of energy (Planck's constant). Einstein further stated that light of frequency (f) could be considered as a stream of photons.

For our experiment, the term's quanta, Planck's constant, boxcars and game marbles will be considered as equivalent. The term marble or boxcar will be used when we discuss our imaginary star. The number of quanta in the blue light from our white star is 7.3×10^{14}. The number of quanta in the green light of our white star is 6.0×10^{14} and for red light 4.3×10^{14}.

These are very large numbers of quanta and very small amounts of energy, therefore, let us reduce them to 7 marbles for blue, 6 marbles for green and 4 marbles for red light when we discuss our imaginary white star. In other words, a train of 7 marbles or boxcars is a blue photon, 6 marbles or boxcars is a green photon and 4 marbles or boxcars is a red photon.

Placing the white star in space at a distance equivalent to our sun's distance, there would be 7 marbles striking the blue receptors in our imaginary eye, 6 striking the green receptors and 4 striking the red receptors every second. When this happens we see white light.

The intensity of the white light is determined by the number of blue, green and red receptors struck, since our eye has many of each of the receptors. These marbles are lined up in a row like boxcars on a train so that each row of marbles is detected by one sensor.

A train of marbles will be the number of marbles line up for one second. Many trains will be traveling through space to the sensors in our eye and the intensity of the light will be determined by how many trains are grouped together in one-second intervals. The color will be determined by how many marbles are in each train.

So in near space our white star looks white to us and the intensity of the white light is determined by how close the star is to us in space. The trains, because of the square of the distance law, strike fewer sensors the farther away our imaginary star is.

We must remember that the intensity of light is determined by the number of trains striking our eye per second and the color is determined by the number of boxcars (marbles) in each train.

The last sentence above is very important to remember. It is also important that the reader understand that the human eye interprets light and colors according to the energy (number of boxcars in a train) and the number of trains striking the eye (intensity) just as it is described herein.

CHARACTER OF DEEP SPACE

In space there are so called quasi-particles permeating free space as defined by quantum mechanics. These quasi-particles blink in and out of existence and probably arise from the vector addition of all the radiation, which fills space emanating from all directions.

We also find charged particles, such as electrons and protons, floating in free space. Our own sun can be seen to emit vast quantities of such particles from its corona. These particles exist in vast numbers emitted by the sun and travel as far as 200 times the earth-sun (au) distance. It is these charged particles and quasi-particles, which give free space impedance (resistance) to the transit of light, and limits its speed to $3x10^8$ meters per second in free space.

Light has different speeds in different mediums, for example, in glass light travels at $2x10^8$ and in an Einstein-Bose condensate of sodium light travels at the unbelievable slow speed of 48 miles per hour (no misprint). A further discussion of light speed is not necessary in this writing.

We should say, however, that accepted physics gives space impedance of 377 ohms. I have calculated that some of that impedance is reactive and some resistance. Since only resistance dissipates energy and my calculations show 4.55ohm resistance for free space, I may use it to show the color change for light energy as it transits space.

A DISCUSSION OF THE EFFECTS OF PARTICLES, QUASI-PARTICLES AND THE RESISTANCE OF FREE SPACE ON LIGHT AS IT TRANSITS FREE SPACE

Compton gave us the equations showing that electrons and other charged particles scatter x-rays. The average scattering angle for all collisions is 45 degrees and Compton's equations show that the wavelength (frequency) change for this angle is 0.200710 nm for an x-ray of 0.20nm. Hence, the fraction of energy lost is 0.00354. These equations can equally be applied to the study of quanta loss from light as it transits free space.

Compton's equations tell us that as our train of marbles collides with charged particles in free space a marble is also knocked out of the train. Remember a train consists of a defined number of marbles passing a given point for duration of one second.

Therefore, if the charged particle is in the path of a blue train then one marble is knocked out of the train. As the train continues to transit free space more and more marbles are knocked out of the train by collisions with charged particles and quasi-particles.

Finally, after 1,000 collisions, the train of light (originally blue) now contains only 6,000 marbles and the green receptor detects green light. There are no trains now with 7,000 marbles and the blue sensors detect nothing.

On average, what was originally blue, green or red will have the same number of marble knocked out of their train. For simplicity of discussion we will follow only the blue light until it turns red.

As the train of marbles continues to transit free space, more and more charged particles knock marbles from the train until the train only contains 4 marbles. The eye now sees a red star.

At this point, the blue and green sensors of the eye detect nothing. The red sensor detects the 4,000 marbles as a red light (originally blue light). This is now the only sensor working.

All other trains have passed out of the visible spectrum and into the infrared, microwave or radio spectrum. If we placed a sensor in their path to detect them, they would be seen as infrared, microwave or radio energy.

The stellar wind, which consists of a vast number of charged particles radiating away from the sun, offers a resistance to the transit of light through space.

Light exerts pressure on these particles as the train of light leaves the sun. This pressure drives the particles farther from the sun. Indeed, newborn stars clear the space surrounding them after birth in dense areas such as the Horsehead Nebula.

Light pressure on mass (matter) causes the light to lose energy as discussed previously regarding the Compton effect. And again, we must say that the effect of energy loss is a change in color of visible light moving it toward the red region of vision. When light loses enough energy to move it out of the visible spectrum, it becomes infrared, then microwave and finally radio waves.

THE COSMOLOGICAL CONSTANT

Scholars are now vigorously debating whether the cosmological constant allows our universe to be 13 billion years old or 17 billion years old. I believe I have presented enough evidence to lay the cosmological constant in the trash can where it belongs and accept the fact that the universe is not expanding but rather that space offers a resistance to the transit of light (all electromagnetic radiation) through space.

Since free space contains particles and quasi-particles, a resistance to the transit of light through space occurs. That resistance has a value of approximately 4.55 ohms and the loss of energy (marbles) as it transits space for one second can be calculated using the equation: $\Delta e = (hf)^2 (4.55)$

Where:

Δe = change of energy in an electromagnet wave in one second

H = Planck's constant

F = frequency of the electromagnetic wave

4.55 = resistance to the transit of an electromagnetic wave

Using the above equation will show that blue light changes to infrared after traveling for 12 billion light years through space. In 1993 we could see less than 1 billion light years and now new instruments allow us to see 12 billion light years.

The Chandra x-ray observatory just now seeing its first light will enable us to see the cosmos in much greater clarity. The normal galaxies containing energies similar to our Milky Way appear in the infrared at this distance due to the resistance offered by particles and quasi-particles in free space.

THE MICROWAVE (3K) BACKGROUND

The microwave background is listed separately in this writing because the advocates of the big bang theory list it as one of the four cornerstones to rest their theory on. It is apparent, if we examine the above discussion, that the microwave background comes mainly from (Milky Way-like) stellar objects in space just beyond the 12 billion light year distance because visible light has changed to the microwave region by the time it gets here from that distant region.

In 1965, Penzias and Wilson detected a microwave radiation background coming from all regions of space at 7.35cm and received a Nobel Prize for their discovery. Proponents of the Big Bang theory proclaimed this discovery as proof of the residual temperature left over from the Big Bang and made it one of the cornerstones of that theory.

WEDDINGS AND PICKET FENCES

You don't have to be Einstein; you don't have to know calculus; you don't even have to know physics to understand why the sky is dark at night, in an infinite universe. If you have an imagination and have been to a wedding or seen a picket fence you can easily see why the sky is dark at night with the following explanation.

If you ever saw a bride with the bridal veil over her face, you would notice that her beauty is enhanced because the lines of her face are softened and the light coming from her face to your eye is reduced. Less light comes from her face to your eye because the veil is between her face and your eye.

The veil is made of holes (empty space) and cloth (matter). Light travels from her face through empty space to your eye but the matter between her face and your eye blocks light. You, therefore, cannot see that portion of her face that is blocked by matter.

A picket fence gives the same effect to what the eye can see as the bridal veil. You can only see objects through the empty space between the pickets. In space there are many bridal veils and picket fences. The reader need only consult some astronomy books to learn that space is filled with matter.

In some places matter is so dense we cannot see stars within or on the other side. For example, the galactic center of our own Milky Way is very hard to see even today with modern instruments because of the dense matter in that region. Even the space between galaxies contains matter.

In an infinite universe there would be an infinite amount of matter between the observer and stars at great distances. This was why stars appear red shifted, as the observer looks deeper and deeper into space. Ordinary stars and galaxies, such as our sun and Milky Way, are visible out to 12 billion light years and then pass into the infrared, microwave and radio spectrum.

ON THE DISTRIBUTION OF THE ELEMENTS IN THE UNIVERSE

Physicists have long thought that the universe was young having an age of less than 20 billion years. The predominance of hydrogen as the major constituent of all the elements found in the universe was a powerful influence in their thinking. If the universe extended infinitely into the past, then the majority of the hydrogen should have condensed into stars.

In turn, the stars convert this basic element into all the other elements. Helium, oxygen, silicon, carbon and finally iron are produced in the fusion process. The cosmos we observe today consists of 99% hydrogen. The majority of the 1% remainder is helium, leaving a very small percentage for all the other elements.

The amount of each element produced and then subsequently fused into other elements depends on the size and life history of the star. Elements heavier than iron are created when the star dies in a sudden cataclysmic explosion. Physicists think that an infinitely aged universe would have little hydrogen and a much larger percentage of helium, oxygen, carbon, silicon and iron.

At the time this thinking was formulated and widely adopted by the scientific community there was little reason to reject the idea. It seemed a reasonable hypothesis and hard to argue against. The hypothesis has now congealed and hardened among virtually all scientists.

In the early 1980's I happened to see a photograph of two stellar objects called 3C273 and M87 which were suspected of containing black holes. Since I had never fully let go of a childhood belief that the universe is infinitely aged, I wondered if these objects might not explain the distribution of the elements in such a universe.

The hypothesis can be made as follows:

Just as the stars accrete hydrogen and generate the other heavier elements, then black holes accrete all elements where they are reduced to their basic constituents of electrons and protons by the immense gravitational and magnetic forces therein. The temperatures generated in the powerful gravitational and magnetic forces near a black hole overwhelm the strong force itself. The extreme temperatures reduce all constituents generated in stars, i.e. all elements in the periodic table, back to electrons and protons.

In some cases, perhaps in 3C273 or some other stellar-like objects such as quasars these basic constituents (electrons, protons) are ejected out the polar-regions in radio jets. The reader should read the July 1993 issue of *Scientific American* for further details on these phenomena.

After traveling some finite distance (200,000 light years) from the stellar object, these electrons and protons lose momentum and congeal again into hydrogen and then into new stars.

Indeed, I believe astronomers are documenting these phenomena in various regions of space.

This phenomenon allows a full circle for the destruction of the heavier elements and their recreation in space as hydrogen and helium. It fully explains the distribution of the elements in an infinitely aged universe.

CONCLUSION

I believe the reader will conclude that we have successfully destroyed three of the cornerstones that the Big Bang theory rests on by using accepted principles and laws in the physics of our time. They were the Doppler red shift, microwave radiation background (3k), and the distribution of the elements.

An alternative explanation was offered using accepted principles and laws of physics. My hope is that the reader will learn the truth from these writings and, ultimately, convey the truth to the most learned among us for more study and evaluation.

CREDITS

The physics discussed in this article was acquired by me over a long lifetime of reading and I no longer remember where I read this or that. I would like to refer the reader to some 1993 *Scientific American* magazine articles, which had a big impact on my thinking. They are: May, "Inconstant Cosmos", Corey S. Powell; "The Most Distant Radio Galaxies", George K. Miley and Kenneth C. Chambers; July, "Edwin Hubble and the Expanding Universe", Donald E. Osterbrock, Joel A. Gwinn and Ronald S. Brashear.

Most of my recent reading comes from, "Physics for Scientists and Engineers with Modern Physics", third edition, volume 2 by Serway and I give this book some of the credit for information contained in this article.

I am grateful to a friend, Marv Sass, who wrote the C-language program enabling me to solve my equation on a personal computer by using the concepts of the calculus. The graph plots the change in energy of blue light (7.3×10^{14}) as it transits free space from our sun out to 50 billion light years and microwave energy (7.35m) detected by Penzias and Wilson just beyond 12 billion light years.

Article 3. FORECASTING THE MOON BY OBSERVATION
By Carrel W. Uptergrove

Nearly everyone knows that the earth and moon revolve around each other or more exactly around a common center of gravity and that the two together revolve around the sun.

Most people living a busy earthbound life rarely pause to observe these motions. People are synchronized to the sun's motion. Most of us get out of bed somewhere close to sunup, dress, eat and go to work.

Then shortly after sundown, we may go to bed and repeat the cycle the following day. If the clock says noon, we know the sun is approximately overhead and when it says midnight, the sun is directly beneath our feet on the other side of the earth.

Occasionally in our busy schedule, we look up and see the moon. We see it sometimes in daylight, sometimes at night. We see its' phases. We wonder when it will be full or new. But these casual observations make the motions of earth, sun, future position or phase will be. It is not necessary to refer to these aids at all and you can learn to predict the phases and motion of moon easily by the following method. We need to know a few facts and conditions first:

Condition 1. We will initially observe the moon at sundown ONLY. By doing this, we have stopped the daily rotation of the earth in the same way that a strobe light, synchronized to the rotation of a wheel, stops the wheel's motion to the eye.

Auto mechanics use a strobe to time your car. If you have never observed a strobe light in operation, you should ask your mechanic to demonstrate the phenomenon.

Fact 1. The full moon always rises at sundown and the new moon always sets at sundown

Fact 2. From an observational viewpoint, the moon goes around the earth in twenty-eight days.

Fact 3. The moon travels from west to east. Your mind may reject this statement at first because you know that the moon travels from east to west, but, remember that this motion is a function of the earth's daily rotation and we are stopping this rotation with our strobe effect. That is, we are observing the moon only at sundown.

You now have all the facts and conditions in your mind. The only thing left to do is to observe the moon at sundown. Let's assume that on the first sundown, we look up and observe the moon exactly on the western horizon. From the facts and conditions just stated, we know that it will always be a new moon when observed at sundown in this position, and the earth, sun and moon will always be aligned in a row with the moon between earth and sun so that the dark side the moon faces the earth.

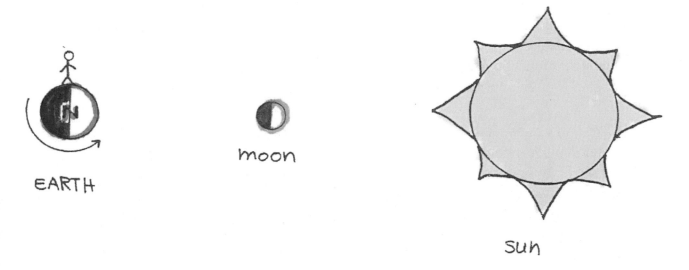

Figure 1. Moons Position

Suppose that our busy schedule does not allow us to observe the moon for three sunsets or perhaps a cloud cover obscures the sky for three days and exactly at sundown, a break in the clouds permits us to view the moon.

Because it takes fourteen days for the moon to travel from the western horizon to the eastern horizon, we will see a crescent moon located 3/14ths of the distance into the visible sky arc to the eastern horizon.

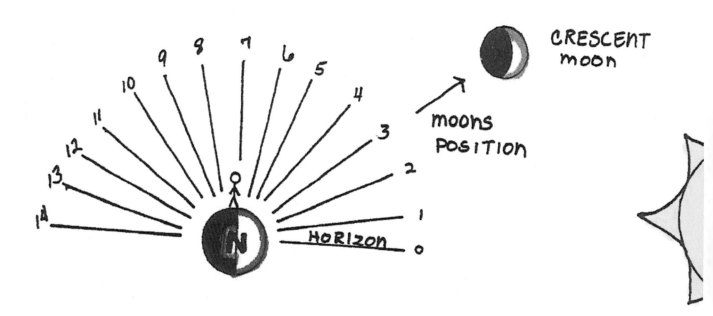

Figure 2. Moons Position

Now, we go about our busy earthly schedules and four days later, we happen to have time to look for the moon at sun down. We observe a half moon directly overhead. We ALWAYS observe a half moon directly overhead at sundown.

CRESCENT

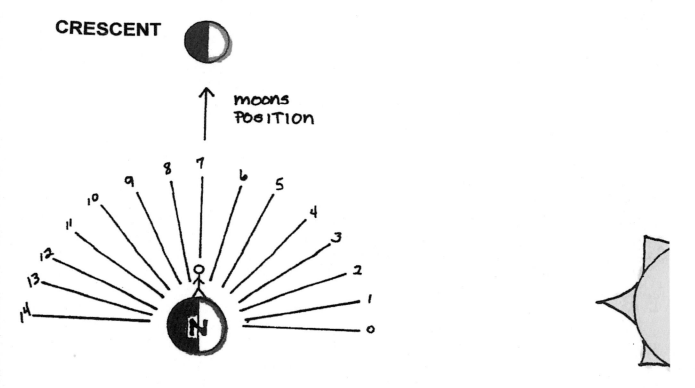

moons
POSITION

8 7
9   6
10   5
11   4
12   3
13   2
14   1
N   0

Figure 3. Moons Position

Again, we go about our schedules and seven days later we happen to observe the full moon on the eastern horizon at sun down.

Figure 4. Moons Position

Of course, we expected this by now and have learned to predict the future position and phase of the moon.

We know now by reasoning that in seven more days the moon will be directly beneath our feet at sundown and therefore is six hours away from being visible on the eastern horizon and will be visible during the morning hours while we are at work or play.

We no longer have to look at the moon only at sundown. Now that we understand its motion and phases, we can look at it anytime it is visible, move it mentally back to sundown, predict its phases, rising and setting and so forth.

For instance, suppose we observe the moon at midnight, directly overhead and we know that sunset occurred at 6 pm. We mentally move the moon back to the 6 p.m. position and see that it rose approximately one hour before sundown.

By observation of the moon and using the facts we have set forth here, we will always be able to understand where the moon is going, where it will be in the future and what is phases will be.

The fact that the earth is tilted on its axis 22 ½ degrees and that earth and

Moon both rotate around the sun could make this discussion more technical.

However, the concept as presented in this discussion should enable the average person to predict the movement of the moon through the sky by casual and random observation.

This ability to predict and understand the moon's motion and phases makes us feel more comfortable in the universe in which we live.

Article 4. FIRE ABOVE THE SUN
By Carrel W. Uptergrove

An article in Science News, "Fire Above the Sun" in the August 31, 1996 issue, stated, "...in the sun's corona, hydrogen reaches a temperature of 6 million Kelvin's and oxygen an astounding 100 million Kelvin's.

Though the sun's corona lies several hundred thousand kilometers above the sun's surface, the corona has a temperature of at least 1 million Kelvin's, about 200 times greater than that of the surface itself.

It's as if the air high above a candle was hotter than the flame that heats it. Exactly how the sun has managed to accomplish such a feat continues to confound astronomers. These numbers indicate that the atoms are heated roughly in proportion to their atomic weights, oxygen being 16 times more massive than hydrogen."

It would be more accurate to say that the atoms are heated roughly in proportion to their ionization state and the effects of ionization over time in a magnetic field. At the temperatures in the sun's corona the ionization state (q) can easily be imagined as changing state from zero to one for hydrogen and zero to eight for oxygen. However, it would seem that very little time would be spent in a low ionization state.

At these high temperatures hydrogen and oxygen would probably be fully ionized virtually at all times. The mean or average ionization state (q) can be calculated from the spectral lines of emission from these elements.

From these measurements the force (F) can be calculated for hydrogen and oxygen. Without those measurements at my hand it seems reasonable to assume that oxygen, on average, will have eight times the charge of hydrogen at these high temperatures, since oxygen has eight electrons and hydrogen has one.

The force (F) exerted on a charged particle follows the equation:

$$F = qvXB \ sin \ \Theta$$

Where:

F = Force applied to the charged particle

Q = Charge of Particle

B = Magnetic Flux Density

V = Velocity of the charged particle

$Sin \ \Theta$ = Angle of velocity vector to the magnetic field

As hydrogen and oxygen atoms leave the dense photosphere of the sun they enter a much more rarified area in the photosphere. Velocity would not have a tendency to decrease, as it does in the corona, by virtue of multidirectional collisions to slow the atoms down because of the increased spacing between atoms; therefore, temperature would not be decreased by these collisions.

Multidirectional collisions have some cooling effect on gas or plasma because the velocities are added by vectoring. For example, some of the atoms can have equal velocity for travel in opposite directions and on a collision course.

When this collision happens all momentum is lost by both atoms and they are no longer in motion. With no motion their temperatures are cooled dramatically and now absorb heat from the surrounding plasma and are again set into motion.

The directional vector inherent in the magnetic flux, driving the atoms away from the photosphere also reduces multidirectional collisions because charges in this region follow the magnetic flux lines away from the sun. This unidirectional flow of particles away from the sun, following the magnetic flux lines greatly enhances the temperature effect due to the reduced collisions.

It's like a farmer who gets all of his horses harnessed and pulling in the same direction. They can go much faster than running in many directions in the barnyard, colliding with each other on occasion.

Note that the force (F) equation is a statement regarding the force created or applied in the 1st second. Therefore, the 2nd second and all subsequent seconds that the atoms are under the influence of the magnetic flux (B) must be accounted for by adding time to the equation.

It is obvious that the force applied in the 1st second increases the velocity (v) of the particle. It is also obvious that the force (F) in the 2nd second is greater than for the 1st second ...*Ad infinitum.*

Accounting for the subsequent seconds that the charge is in the corona each particle has a total force (F_t).

$$F_t = F_1 + F_2 + F_3...etc.$$

Where:

F_1 = force applied in the 1st second

F_2 = force applied in the 2nd second

F_3 = force applied in the 3rd second ... etc.

And the velocity changes because of the forces applied to the particles as they follow the magnetic flux lines away from the corona, so that:

v1 is less than v2 is less than v3 ... etc.
v1 <v2 <v3 ...etc.

Therefore, in the case of hydrogen for the 1st second:

$F = qvB \sin \Theta$

And for oxygen assuming an average charge of eight (8) times that of the average for oxygen:

$F = 8qvB \sin \Theta$

This is a better explanation of the different temperatures for hydrogen and oxygen than some explanation using weight as an influencing factor. Such an argument would be similar to saying that the electron orbits the hydrogen proton because both have mass and gravitationally attract each other and opposite charges also attract.

The following shows the error of citing gravity:

Given:

F_e = force on a charged particle

k = coulomb constant = 9.0×10^9

e = elementary charge = 1.6×10^{-19}

r = Bohr radius = 5.3×10^{-11}

F_g = gravitational force

G = gravitational constant = 6.7×10^{-11}

M_e = mass of electron = 9.11×10^{-31} kg

M_p = mass of proton = 1.67×10^{-27} kg

For electrical attraction the equation is:

$F_e = k_e^2/r^2 = 8.2 \times 10^{-8}$ Newton's

For gravity:

$F_g = Gm_em_p/r^2 = 3.6\text{x}10^{-47}$ Newton's

The ratio is:

$F_e/F_g = 3\text{x}10^{-39}$

Thus gravitational attraction can be discounted in comparison to the tremendous electrical attraction in the hydrogen atom. Weight and gravity can be discounted in explaining the temperatures of hydrogen and oxygen in the sun's corona.

It would be similar to saying that an apple is smaller because one water molecule evaporated away as it lay on the counter.

In summary, the high temperatures in the sun's corona are caused by:

1. The initial velocity of the charged particle and the velocity change during the time it is in the influence of the magnetic flux lines of the corona.

2. The magnitude of the magnetic flux lines (B).

3. The magnitude of qv.

4. Sin ☉.

5. A rarefied environment in the corona.

6. A reduced collision rate due to a directional flow away from the sun and along the magnetic flux lines.

Article 5. FRESH WATER
By Carrel W. Uptergrove

There are 4 forces in the universe. They are the strong force, weak force, electromagnetic, and gravity. Thanks to Tesla, we use the electromagnetic force extensively. Unfortunately, this force uses fossil fuels to generate power except in places like Niagara Falls or the Hoover Dam.

I propose a way to generate fresh water at sea level from sea water using gravity with a dramatic reduction of the use of fossil fuels. The process is simple and straightforward as follows. Insert a pipe into the sea with a reverse osmosis membrane at a depth which will force water through the membrane.

The sea pressure is 15 pounds per square inch for each 30 feet the pipe is submerged. So, inserted at the proper depth, sea pressure will force fresh water to appear in the pipe.

Next install, probably on the sea floor, a tank that would be much like an oil storage tank. Modified to work like an accordion, with a heavy weight on top and a closed storage area for air when desired.

With a pipeline connected from the fresh water area above the membrane to the bottom of the tank and the tank empty of air except for the closed air area at the top of the tank. Note: the closed air area in the tank makes the tank slightly buoyant even with the weight on top.

Opening a valve between fresh water above the membrane and the tank will cause the fresh water to flow to the tank because of the slight buoyancy caused by the air in the closed area at the top of the tank. Water will flow until the tank is full because it is slightly buoyant. This process opens the accordion tank to the full height.

With a pipeline, attached with a valve to the bottom of the tank, the water in the tank will now flow through the pipe to the surface of the sea when a valve is opened allowing the air in closed storage area to flow out into the sea and the valve at the bottom of the tank is also opened.

This flow out of the tank happens because of the weight at the top of the tank collapsing the accordion.

The cycle can now be repeated by pumping air back into the closed area for air. It is that simple.

Article 6. QUASARS AND BLACK HOLES
By Carrel W. Uptergrove

Today quasars are surrounded in controversy among cosmologists. Some contend that quasars are in the farthest reaches of the universe. Indeed, are the most distant known objects in the universe. They further contend that quasars are receding even farther from us at speeds approaching 1/10 to ½ the speed of light and that this is manifested by a Doppler shift of the spectral energy lines reaching us.

Others claim that quasars would have to contain 100 million solar masses to generate the energy required for the observed intensities at these distances. Therefore, they contend that quasars are in near space.

In short, quasars are enigmatic objects because of their apparent great distances from us, their apparently great energy requirements, their apparent speed of recession from us and their small size.

In this discussion, we propose to build a model which will explain these anomalies regarding quasars. In order to do this, we must first review some basic ideas regarding the universe.

TIME AND SPACE

This model assumes that time extends infinitely into the past as well as into the future and that space extends to infinity in all of the infinite direction around us. While infinity of time and space are difficult to comprehend, it is much easier to accept than try to explain what comes after either or both come to an end. It follows then that we live in a very large universe.

BLACK HOLES

The generally accepted definition of a black hole is described in this manner: A region of space from which, according to the general theory of relativity, neither radiation nor matter can escape.

The matter and energy contained in the black hole collapses toward the center forever. This is a good beginning in the attempt to define and visualize a black hole but it is only a small beginning.

In our model, black holes will have a broader definition and will behave in different ways depending on their size and their environment in surrounding space. Black holes can be very small or very large. In a relative way, they can be hot or cool.

They even spew forth energy and matter in defiance of their definition. They may be stable in size or they may be enlarging by absorbing matter from the surrounding space.

QUASARS

To explain the apparent Doppler shift of the radiation reaching us from the quasars in the farthest reaches of space, we need to look at the basic nature of light.

Einstein theorized that light had mass as well as wavelike characteristics and this has been proven in many experiments through the years.

According to the theory of relativity, spectral lines are redshifted slightly when they leave a star or any massive body. The more massive the body, the more redshift occurs. We can then say that the gravitational attraction of light to a massive body changes the frequency of light slightly.

Since light by definition cannot change speed but is always constant, the gravitation attraction of light to the massive body it is leaving, manifests itself by a redshift. The redshift is directly proportional to the mass of the body and distance to the observer.

Given:

    M = mass

    D = distance separating observer and massive body

    z = redshift

    & = proportional to

    Then: z = Md

This equation clearly demonstrates that a massive body and great distance can significantly redshift an observed object.

This equation can be graphically demonstrated using Planck curves for black bodies at different temperatures.

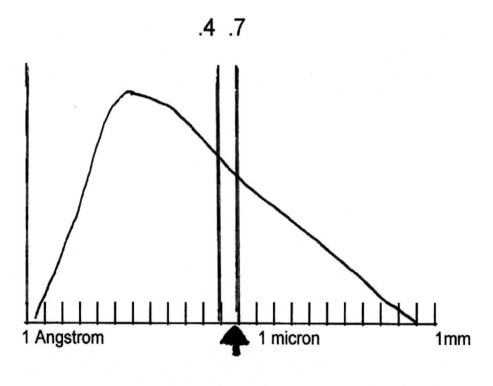

Figure 1. Shows a Planck curve for radiation from a
small black body at a moderate distance.

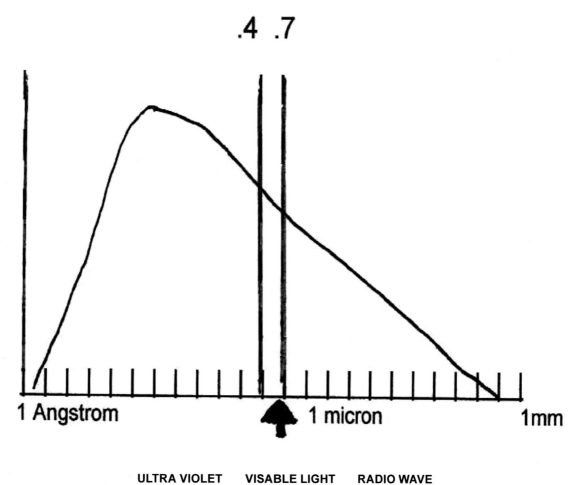

.4 .7

1 Angstrom   1 micron   1mm

ULTRA VIOLET   VISABLE LIGHT   RADIO WAVE

Figure 2. Shows a Planck curve for radiation from a massive
black body at the same moderate distance.

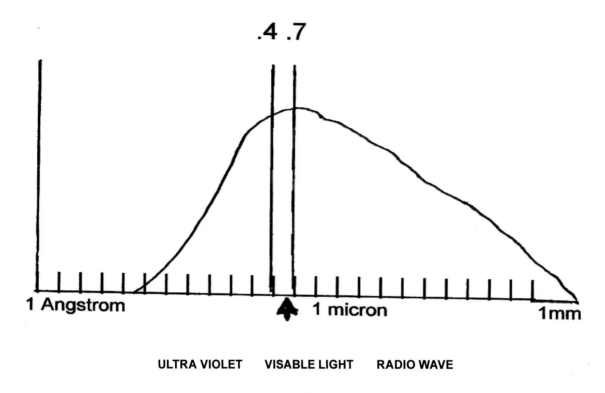

.4 .7

1 Angstrom   1 micron   1mm

ULTRA VIOLET   VISABLE LIGHT   RADIO WAVE

Figure 3. Shows a Planck curve for a massive black body from a great distance.

The significant difference between these figures is that the entire energy spectrum is shifted to the lower frequency end of Figure 2, by the greater mass of the black body and shifted even further in Figure 3, by the combined influence of greater mass, plus greater distance.

On a comparative basis we notice that radio emissions are increased and ultraviolet and light emissions are decreased. This shift in the Planck curves explains one reason why quasars are strong emitters of radio waves.

From this discussion, we can carry the logic one step further and visualize a black body so massive and at such a great distance from the observer as to place the Planck curve outside the visible region of the spectrum as illustrated by Figure 4.

42

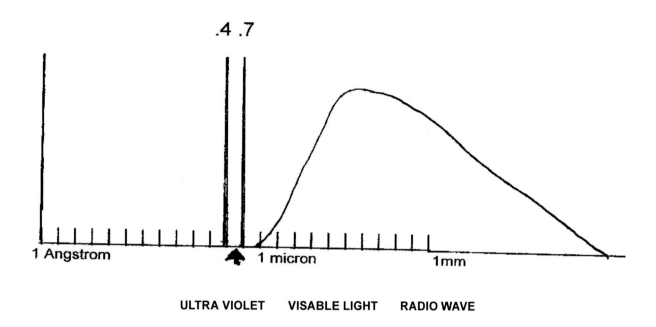

.4 .7

1 Angstrom   1 micron   1mm

ULTRA VIOLET   VISABLE LIGHT   RADIO WAVE

Figure 4. At this distance, the quasar ceases to be a visible object in the sky and becomes an invisible radio source.

THE GREEN LIGHT TURNS AMBER AND THEN RED

Since radiation leaves the quasar at a constant energy level (the speed of light) regardless of the size or temperature of the source there is a finite distance beyond which radiation cannot travel. If the quasar has greater mass, the intensity of radiation increases but the speed of departure remains constant.

The mass of any single photon is influenced to slowdown even more in any single unit of time (1 second). Since red shifts are observed in massive stars as close as 100 light years, it is easy to see that the frequency of light can be considerably altered as it leaves a body as massive as a quasar and as distant as 8 to 20 billion light years.

Photons leaving a gravity field are slowed by that field. The slowdown is inversely proportional to the mass of the body it is leaving.

Given:

S = speed of photon

M = mass of source

& = is proportional to

Then: S & 1/M

The slowdown of the photon is further inversely proportional to the time it is influenced by the gravity field.

Given:

t = time

Then: S & 1/Mt

The above equation says that the slowdown of the speed of departure of the photon is manifested as a frequency change (red shift) rather than a change of speed of the radiation. The green light turns amber, the amber light turns red, and the red light leaves the visible spectrum and is no longer defined as light. This obeys Einstein's general theory of relativity.

WATCHING THE WAVES

Another way to look at the effect which gravity has on light (radiation) is as a wave rather than as a particle with mass. To do this, imagine an observer in space watching the waves go by.

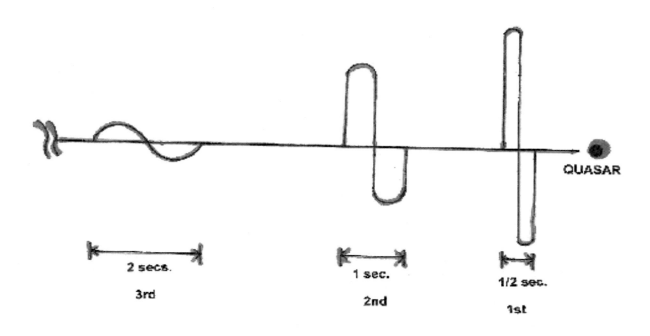

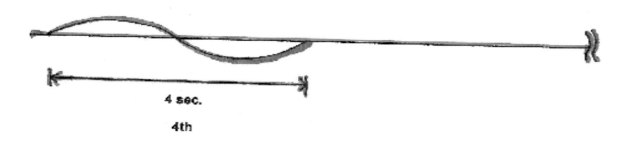

Figure 5. Look at the effect which gravity has on light.
Imagine you are watching waves go by.

At Position 1, the observer notices radiation (waves) leaving the massive body as a Planck curve for black bodies but slightly red shifted due to the great mass of the body.

Looking at a single wave rather than the entire spectrum, we might observe a wave with the amplitude and frequency as at Position 1.

For convenience, let us say that it took ½ seconds for the wave to pass at this point.

The next observer at Position 2 and several billion light years away observes the same wave and due to the effects of gravity and time (Mt) observes that twice as much time is required for the wave to pass him.

Likewise, the observers at Position 3 and 4 notice an increase in time required for the wave to pass.

But, since frequency is inversely proportional to time, we must conclude that the wave has changed frequency.

Given:

f = frequency

Then: f & $1/t$

If the observer at Position 1 had chosen to measure a wave of green light then the observer at Position 2, viewing the same wave would say it was amber and the observer at Position 3 would say it was red.

ENERGY OUTPUT OF QUASARS

Now that we have discussed the red shift of quasars, even though their relative motion may be stationary in respect to us, due to their great mass and great distance, we are left with the problem of explaining the massive energy output required to shine with the brilliance they do from such vast distances

One such possible model might be a massive black hole in the center of a massive galaxy. This galaxy would have stars totaling 10 to 1000 billion solar masses. The galaxy would be observed from its polar regions just as the whirlpool galaxy M51 in the constellation *Canes Venatici*. See Figure 6.

Figure 6. Massive black hole in center of massive galaxy.

Galaxies which do not have their polar regions facing the observer would not be visible from these great distances. Some of the reasons the galaxy is more visible from the polar regions is obvious. For example, more stars are visible to the eye in a polar view versus an edge-on view.

However, the primary reason is because all radiating bodies emit more radiation from their polar regions than they do from their equatorial regions as depicted in Figure 7.

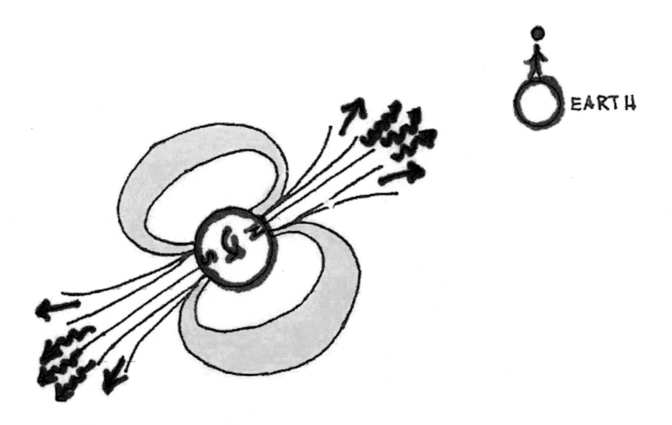

Figure 7. All radiating bodies emit more radiation from polar regions than from equatorial regions.

The more rapidly the body rotates, the more energy is radiated out at the polar regions in comparison to the equatorial region. The sun would be brighter to an observer 93 million miles above its polar region that it is to us and the sun would be brighter still to that observer if the rotation of the sun speeded up, while to the earth it would grow dimmer. Pulsars are a good example of high intensity radiation from polar regions of radiating bodies.

The black hole in the center of the galaxy of our model has been collapsing now for 5 billion years and as a result of this collapse has enormously increased its speed of rotation obeying the law of "the conservation of angular momentum" just as the ice skater obeys the same law by speeding up her rotation when she pulls in her arms performing a pirouette. See Figure 8.

Figure 8. Skater speeds up rotation.

Due to the enormous speed of rotation, the black hole begins to radiate vast amounts of energy into space from its polar regions. To an observer looking down on its polar region, the black hole would appear as a point of light in the sky, coming from a great distance.

Since the black hole is now radiating prodigious amounts of energy, and therefore mass ($E=MC^{-2}$), its speed of rotation can again increase enormously and there for spew out even greater amounts of matter and energy until it consumes itself by ejecting its entire mass out its polar regions spinning ever faster and faster.

This death of a black hole is indeed possible. However, our black hole's speed of rotation is maintained constant due to the fact that matter is flowing into the hole by accretion from the surround galaxy. See Figure 9.

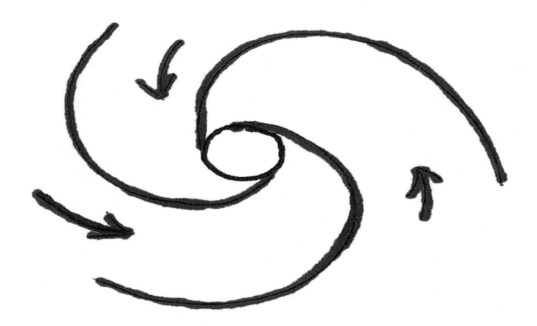

Figure 9. Accretion flow.

This accretion flow is just sufficient to allow the rotation speed of the black hole to remain constant or perhaps increase in speed slightly as the diameter of the galaxy is reduced.

The enormous magnetic flux of the rotating black hole extends into the galaxy and tends to increase the speed of the stars in orbit around the black hole to higher and higher orbital speeds.

This causes a loss of energy by the black hole and is manifested by drag. While the black hole would tend to spin even faster, the drag of the galaxy moderates this increase as the material in the galaxy spirals ever inward toward the black hole.

As the galaxy decreases in diameter over several billion years, the conservation of angular momentum allows an increase in rotation speed of the black hole.

The increased speed of rotation also speeds up the magnetic flux lines dragging the stars ever inward toward the black hole at greater and greater velocities.

Eventually, the galaxy diameter is reduced enough to allow what, has now become an even larger massive black hole, to achieve a fantastic rate of spin due to reduced drag from the galaxy.

A rapidly spinning black hole has two event horizons and the singularity has changed from a point to a circle. See Figure 10.

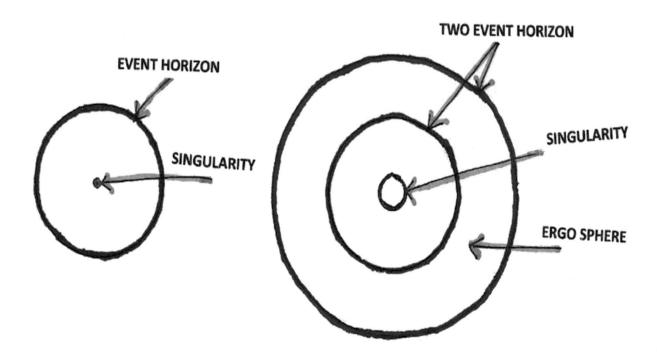

Figure 10. A rapidly spinning black hole.

The area between the two event horizons is called the ergo sphere.

Star clusters entering the ergo sphere can be accelerated to fantastic speeds and ejected out of the black hole in the same manner that an automobile tire can accelerate a rock and eject it into the air.

The exploding galaxy, NGC 5128, is probably an example of this mechanism.

This ejected matter usually occurs in the polar plane of the black hole and its associated galaxy. See Figure 11.

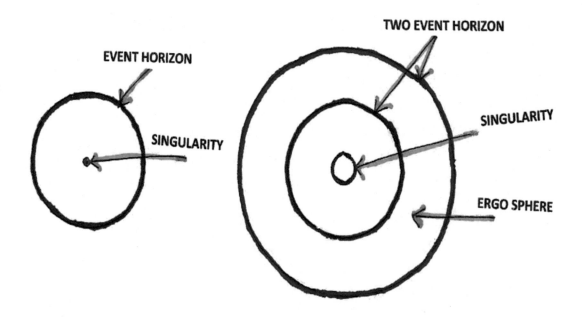

Figure 11. Ejected material usually occurs in equatorial plane.

Matter not ejected by the ergo sphere now falls into and through the singularity. In falling through the singularity, the matter is ejected at great speed out the north and south polar regions of the black hole as a jet of energy and matter. 3C273 and N87 might be such an event. See Figure 12.

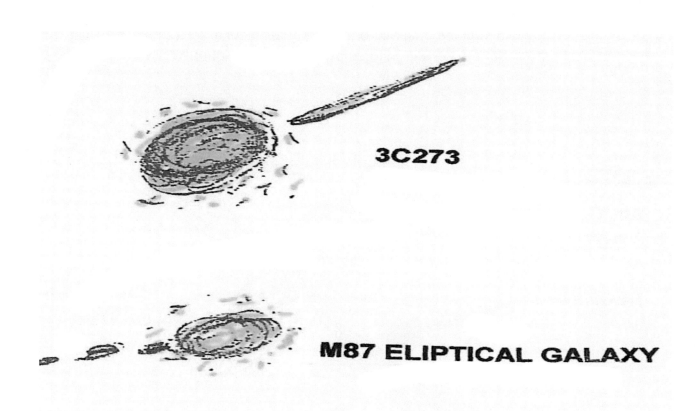

3C273

M87 ELIPTICAL GALAXY

Figure 12. A jet of energy and matter.

The matter in the ejected jets reduces all of the elements, i.e., hydrogen, helium, gold, etc., back to basic constituents such as quarks, neutrons, protons and electrons.

It can now be seen that rapidly spinning black holes is the mechanism to convert matter to energy and eject energy out of its polar regions.

This energy travels vast distances into space, condenses again into matter (galaxies) and begins the cycle over again.

By this mechanism, regions of space once empty are now filled with matter while regions filled with matter are converted to empty space. See Figure 13.

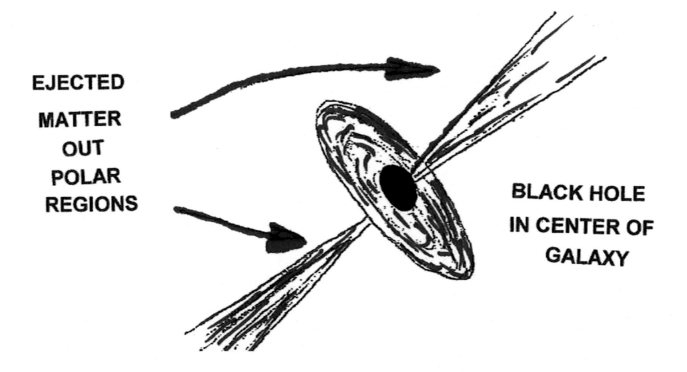

EJECTED
MATTER
OUT
POLAR
REGIONS

BLACK HOLE
IN CENTER OF
GALAXY

Figure 13. The cycle begins again.

Article 7. THE HUBBLE RED SHIFT
By Carrel W. Uptergrove

The Hubble red shift can be explained without reliance on a Doppler shift but rather on the attenuation of light by its transit through the impedance of free space ($\mu_0 c$).

Einstein showed that light has energy (E) and that the energy is proportional to its frequency E = hf, where h is Planck's constant. The theory is that the energy in light is carried in quanta of energy equal to Planck's constant.

Planck's constant is expressed in Joules per second (a quanta of energy) and E in Joules. To understand the implication of these expressions, let us discuss some basic definitions and concepts.

Power lost or dissipated in an impedance is expressed by the equation: P = I^2R. We can then say that the change in energy (ΔE) of a photon can be expressed ΔE = P = 1^2R.

In the case of free space, the impedance is expressed as ($\mu_0 c$).

The definition of current is that if (2 pi) x 10^{18} e, pass a given point in one second, it is one ampere.

This same quantity qualifies as one volt if this magnitude is measured as a difference between two points.

Also this same quantity qualifies as one watt if dissipated in an impedance in one second.

It follows then that if e Joules or h Joules pass a given point in one second, it constitutes an equivalence of a current (I) of that magnitude.

The expression $E = hf$ is the number of quanta of energy (e) passing a given point in one second and therefore qualifies as an equivalent current (I) of energy. Therefore, the expression $P = \Delta E = I^2 R$ can be expressed as equation (1) one.

Equation (1). $\Delta E = (hf)^2 \mu_o c$

It is obvious from equation (1) that after light or any electromagnetic wave has traveled through space for one (1) second, it dissipates energy and therefore, now possesses lower energy (E). It follows then, that it also possesses a lower frequency (f) and longer wavelength.

Where: $E_2 =$ Energy (E) of photon after one second.

$E_1 =$ Initial energy at the beginning of the second being considered.

Then:

Equation (2) $E_2 = E_1 - \Delta E_1$

It follows that as each second of time (T) passes, more energy (ΔE) is lost in its passage through the impedance of free space ($\mu_o c$). Therefore, after a passage of time, the new energy and lower frequency can be expressed as:

$f_1 = 7.31 \times 10^{14}$ hertz $= 410.2$ nm

$E_1 = hf_1 = (6.626 \times 10^{-34})(7.31 \times 10^1) = 4.84 \times 10^{-19}$ J

$\Delta E_1 = (hf_1)^2 (377) = (4.84 \times 10^{-19})^2 (377) = (23.4 \times 10^{-38}(377) = (8.83 \times 10^{-35}$ J

$\qquad = (4.84 \times 10^{-19}) - (8.83 \times 10^{-35})$

$E_2 = E_1 - \Delta E_1 = (hf_1) - [(4.84 \times 10^{-19})^2 (377)]$

$\qquad = (4.84 \times 10^{-19}) - (8.83 \times 10^{-35})$

To simplify the above expressions, where (T) is equal to time, then the value of E after the passage of (3) seconds is:

Time in Seconds

0 sec $E_1 = 4.84 \times 10^{-19}\,J$

             $\Delta E_1 = E_1{}^2 (377)$

1 sec. $E_2 = E_1 - \Delta E_1$

             $\Delta E_2 = E_2{}^2 (377)$

2 sec. $E_3 = E_2 - \Delta E_2$

             $\Delta E_3 = E_3{}^2 (377)$

3 sec. $E_4 = E_3 - \Delta E_3$

If we wish to know the value of energy (E) possessed by the photon after it has traveled for one year, then we must solve the expressions above 31.5×10^6 times because that is how many seconds are in (1) year.

Figure 1 shows a Hubble expansion drawing. The vertical column of a Hubble drawing is usually labeled "velocity", "kilometers/second of recession". The horizontal row is usually labeled "Distance", "Parsecs".

This drawing is now labeled "f, E of 410.2 nm wavelength as seen by observer on earth". The horizontal row is labeled "Distance", "Billions of Light Years". The diagonal line is labeled "$\mu_o c = 377$ ohm = impedance of free space".

The junction of the three lines in the lower left hand corner represents an observer on earth with an energy source E of 410.2 nm. As labeled on the drawing, an observer looking at this energy source sees a frequency of $7.31 \times 10^{-14}h$ and measures energy E of 4.84×10^{-19} J.

If this same energy source could now be moved into space to a location in Ursa Major, then the observer on earth would see a frequency (f) of 6.94×10^{14} hertz and measure energy E of 4.6×10^{19} E. Ursa Major is located 98×10^6 light years from earth which represents:

$(31.5 \times 10^6)\,(98 \times 10^6) = 3.1 \times 10^{15}$ seconds

Therefore, if equations (1) and (2) were solved 3.1 x 10 seconds, the result should yield the same results as given in Figure 1 for Ursa Major. For 410.2 nm wave length emitted from Bootes and Hydra, an earth based observer will see 470 nm wave length and 513 nm wave length respectively.

In summary, all electromagnetic waves lose energy (E) as they transit free space and therefore frequency (f) is reduced according to equation: $f = E/h$.

Fundamental Constants

$$\mu_o c = 4 \times 3.1416 \times 10^{-7} \text{ N/A}^2$$

$$c = 2.997 \times 10^8 \text{ m/s}$$

$$h = 6.626 \times 10^{-34} \text{ J.s}$$

$$e = 1.602 \times 10^{-19} \text{ J}$$

The impedance of free space is given as $\mu_o c$ being equal to 377 ohms. The definition of impedance is that it is "the vector sum of (Xl) inductive reactance, (Xc) capacitive reactance and R resistance". No power is consumed by any impedance other than R resistance. Therefore, equation (1) must modify the expression $\mu_o c$ so that it expresses only the resistance (R).

If observations by the theorists on the Hubble red shift reveal that the power lost in $\mu_o c$ is approximately one tenth (1/10) of that computed by equation (1), then $\mu_o c$ must contain a reactance and a resistance. Angle θ expresses the vector relationship between $\mu_o c$ and R. Figure 2 shows this relationship graphically.

We can then say that: $\cos \theta = R / \mu_o c - 0.1$

Assuming that R is approximately 1/10th of $\mu_o c$, Equation 1 can now be rewritten as Equation 3.

Equation (3)

$E = (hf)^2 (\mu_o c) (R / \mu_o c)$ And reduces to: $E = (hf)^2 (37.7)$

If the R resistive component of the $\mu_O c$ impedance of free space is assumed to be approximately 37.7 ohms and time (T) is divided into a reasonable number of increments, then equation (3) can be solved to a rough approximation. Figure 3 shows the chart of ten incremental calculations out to Ursa Major. The calculations are as follows:

If T = 308.7 x 10^{12} seconds = one tenth of distance to Ursa Major

$$\Delta e_i = E^2{}_i \, (\mu_O c \cos \theta)$$

1st of ten $= (4.84 \times 10^{-19})^2 \, (37.7)$

incremental $= 8.822 \times 10^{-36} \, J$

calculations $\Delta E_i = \Delta e_i (T)$

$= 2.7233 \times 10^{-21}$

$E_1 = E_i - \Delta E_i = (4.84 \times 10^{-19}) - (2.72 \times 10^{-21})$

$= 4.813 \times 10^{-19} \, J$

****

2nd $\Delta e_1 = E_1{}^2 (R) = (4.813 \times 10^{-19})^2 \, (R)$

increment $= 8.73 \times 10^{-36} \, J$

$\Delta E_1 = \Delta e_1 (T) = 2.69 \times 10^{-21}$

$E_2 = E_1 - \Delta E_1 = (4.813 \times 10^{-19}) - (2.695 \times 10^{-21})$

$= 4.786 \times 10^{-19} \, J$

****

3rd $\Delta e_2 = E_2{}^2 (R) = 8.635 \times 10^{-36} \, J$

increment $\Delta E_2 = \Delta e_2 - \Delta E^2 (T) = 2.665 \times 10^{-21}$

$E_3 = E_2 - \Delta E_2 = 4.759 \times 10^{-19}$

****

4th $\Delta e_3 = E_3^2 (R) = 8.54 \times 10^{-36}$

increment $\Delta E_3 = \Delta e_3 (T) = 2.635 \times 10^{-21}$

$E_4 = E_3 - \Delta E_3 = 4.733 \times 10^{-19}$

****

5th $\Delta e_4 = 4_4^2 (R) = 8.555 \times 10^{-36}$

increment $\Delta E_4 = \Delta e_4 (T) = 2.607 \times 10^{-21}$

$E_5 = E_4 - \Delta E_4 = 4.707 \times 10^{-19}$

****

6th $\Delta e_5 = E_5^2 (R) = 8.35 \times 10^{-36}$

increment $\Delta E_5 = \Delta e_5 (T) = 2.578 \times 10^{-21}$

$E_6 = 4_5 - \Delta E_5 = 4.68 \times 10^{-19}$

****

7th $\Delta e_6 = 6_3^2 (R) = 8.26 \times 10^{-36}$

increment $\Delta E_6 = \Delta e_6 (T) = 2.55 \times 10^{-21}$

$E_7 = E_6 - \Delta E_6 = 4.654 \times 10^{-19}$

****

8th $\Delta e_7 = 7_3{}^2\ (R) = 8.167 \times 10^{-36}$

increment $\Delta E_7 = \Delta e_7\ (T) = 2.52 \times 10^{-21}$

$E_8 = E_7 - \Delta E_7 = 4.63 \times 10^{-19}$

****

9th $\Delta e_8 = E_8{}^2\ (R) = 8.08 \times 10^{-36}$

increment $\Delta E_8 = \Delta e_8\ (T) = 2.494 \times 10^{-21}$

$E_9 = E_8 - \Delta 8_8 = 4.604 \times 10^{-19}$

****

10th $\Delta e_9 = E_9{}^2\ (R) = 7.99 \times 10^{-36}$

increment $\Delta E_9 = \Delta e_9\ (T) = 2.467 \times 10^{-21}$

$E_{10} = E_9 - \Delta E_9 = 4.579 \times 10^{-19}$

****

Thus calculations can be made using this method to compute the red shift to Bootes and Hydra or to any object in space.

It should be obvious now that the Big Bang never happened and that the Hubble red shift is not a Doppler effect.

It can be shown with Equation (3) that the microwave background (3K) is the result of the attenuation of higher frequencies to the microwave wavelengths.

Article 8. THE INFINITELY AGED UNIVERSE
By Carrel W. Uptergrove

The Hubble red shift can be explained without reliance on a Doppler shift but rather on the attenuation of light by its transit through the impedance of free space ($\mu_o c$).

Einstein showed that light has energy (E) and that the energy is proportional to its frequency E = hf, where h is Planck's constant. The theory is that the energy in light is carried in quanta of energy equal to Planck's constant.

Planck's constant is expressed in Joules per second (a quanta of energy) and E in Joules. To understand the implication of these expressions, let us discuss some basic definitions and concepts.

Power lost or dissipated in an impedance is expressed by the equation, $P = I^2 R$. We can then say that the change in energy $\Delta E = P = I^2 R$.

In the case of free space, the impedance is expressed as ($\mu_o c$).

The definition of current is that if (2 pi) x 10^{18} e, pass a given point in one second, it is one ampere.

This same quantity qualifies as one volt if this magnitude is measured as a difference between two points.

Also this same quantity qualified as one watt if dissipated in an impedance in one second.

It follows then that *e* Joules or *h* Joules pass a given point in one second, it constitutes an equivalence of a current (I) of that magnitude.

The expression E = hf is the number of quanta of energy (e) passing a given point in one second and therefore qualifies as an equivalent current (I) of energy.

Therefore, the expression P = ΔE = I^2R can be expressed in Equation 1.

Equation 1. ΔE = $(hf)^2 \mu_o c$

It is obvious from Equation 1, that after light or any electromagnetic wave has traveled through space for one (1) second, it dissipates energy and therefore, now possesses a lower energy (E). It follows then, that it also possesses a lower frequency (f) and longer wavelength.

Where: E_2 = Energy (E) of photon after one second.

E_1 = Initial energy at the beginning of the second being considered.

Then:

Equation 2. $E_2 = E_1 - \Delta E_1$

It follows that as each second of time (T) passes, more energy (ΔE) is lost in its passage through the impedance of free space (μ_oc). Therefore, after a passage of time, the new energy and lower frequency can be expressed as:

f_1 = 7.31 x 10^{14} hertz = 410.2 nm

E_1 = hf_1 = (6.626 x 10^{-34}) (7.31 x 10^{-14}) = 4.84 x 10^{-19}) J

ΔE_1 = $(hf_1)^2$ (377) = (4.84 x $10^{-19})^2$ (377) = (23.4 x 10^{-38}) (377) = 8.83 x 10^{-35} J

E_2 = E_1 - ΔE_1 = (hf_1) – [(4.84 x $10^{-19})^2$ (377)] = (4.84 x 10^{-19}) – (8.83 x 10^{-35})

To simplify the above expressions, where (T) is equal to time, then the value of E after the passage of (3) seconds is:

<u>Time in Seconds</u>

0 sec. $E_1 = 4.84 \times 10^{-19}$ J

$$\Delta E_1 = E_1{}^2 (377)$$

1 sec. $E_2 = E_1 - \Delta E_1$

$$\Delta E_2 = E_2{}^2 (377)$$

2 sec. $E_3 = E_2 - \Delta E_2$

$$\Delta E_3 = E_3{}^2 (377)$$

3 sec. $\Delta E_4 = E_3 - \Delta E_3$

If we wish to know the value of energy € possessed by the photon after it has traveled for one year, then we must solve the expressions above 31.5×10^6 times, because that is how many seconds are in (1) year.

Calculations using equation (1) give different results from those actually observed. Therefore, we must assume that the impedance of free space has reactive and resistive components.

The impedance of free space is given as ($\mu_0 c$) being equal to 377 ohms. The definition of impedance is that it is "the vector sum of (**XL**) inductive reactance, (**XC**) capacitive reactance and (R) resistance". No power is consumed by any impedance other than (R) resistance.

Therefore, equation (1) must modify the expression ($\mu_0 c$) so that it expresses only the resistance (R).

If observations by the theorists on the Hubble red shift reveal that the power lost in $\mu_0 c$ is a fraction of that computed by equation (1), then $\mu_0 c$ must contain a reactance and a resistance. Angle θ expresses the vector relationship between

(μ_Oc) and R. See <u>Figure 2.</u>

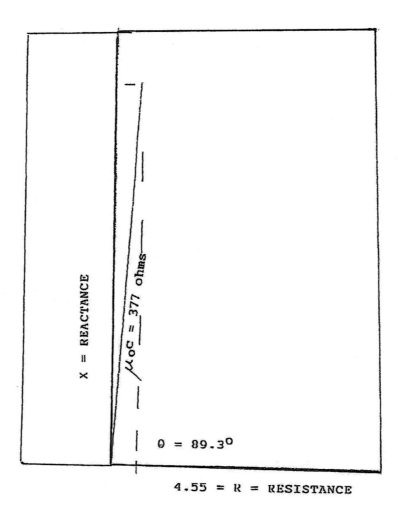

Figure 2. Graphically shows the vector relationship between (μ_Oc) and R.

We can then say that: $\cos \theta = R/\mu_O c = 0.0120689$.

Assuming that R is approximately 4.55 ohms.

Equation 1. can now be rewritten as Equation 3.

Equation 3. $\Delta E = (hf)^2 (\mu_O c) (\cos \theta)$

and reduces to: $\Delta E = (hf)^2 (4.55)$

If the R resistive component of the (μ_0c) impedance of free space is assumed to be approximately 4.55 ohms and time (T) is divided into a reasonable number of increments, then Equation 3 can be solved to a rough approximation.

Thus calculations can be made using this method to compute the red shift to Bootes and Hydra or to any object in space.

Figure 1. Shows a Hubble expansion drawing. The vertical column of a Hubble drawing is usually labeled "velocity" "kilometers/second of recession". The horizontal row is usually "Distance" "Parsecs".

This drawing is now labeled "f, E of 410.2 nm wavelength as seen by observer on earth". The horizontal row is label "Distance", "Billions of Light Years". The diagonal line is labeled "(μ_0c) (cos θ) = 4.55 ohm = impedance of free space".

The junction of the three lines in the lower left hand corner represents an observer on earth with an energy source € of 410.2nm. As labeled on the drawing, an observer looking at this energy source sees a frequency of 7.31×10^{14} h and measures energy € of 4.84×10^{-19} J.

If this same energy source could now be moved into space to a location in Hydra, then the observer on earth would see a frequency (f) of 5.85×10^{14} hertz and measure energy € of 3.88×10^{19} J. Hydra is located 3.59×10^9 energy source light years from earth which represents:

$(31.5 \times 10^6) (3.59 \times 10^9) = 113.1 \times 10^{15}$ seconds

Therefore, if equations (3) and (2) were solved 113.1×10^{15} times, the result should yield the same results as given in Figure 1. For Hydra.

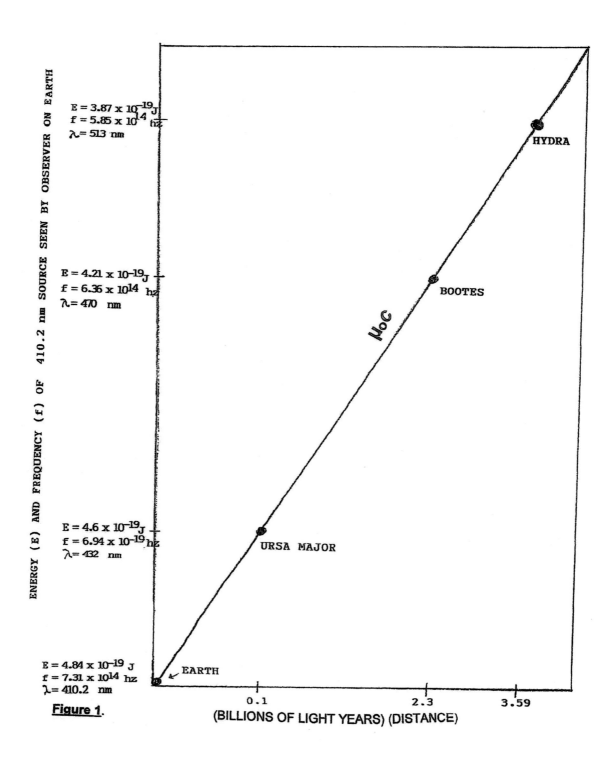

Figure 1. Shows a Hubble expansion drawing.

For 410.2 nm wavelength emitted from Bootes, an earth based observer will see 470 nm. In summary, all electromagnetic waves lose energy € as they transit free space and therefore frequency (f) is reduced according to equation f = F/h.

Fundamental Constants

$$\mu_O = 4 \text{ pi} \times 10^{-7} \text{ N/A}^2$$

$$c = 2.997 \times 10^8 \text{ m/s}$$

$$h = 6.626 \times 10^{-34} \text{ J.s}$$

$$e = 1.602 \times 10^{-19} \text{ J}$$

The Microwave (3K) Background

It can be shown with Equation 3. That the microwave background (3K) is mostly the result of the attenuation of higher frequencies to the microwave wavelengths. See Figure 3.

Figure 3. plots the 410.2 nm wavelength with energy levels from one to five x 10^{-19} J. Figure 3. Extends to fifty billion light years.

One can readily see by means of this figure, that 410 nm (blue light) becomes 550 nm (yellow light) after traveling six (6) billion light years. After traveling twelve (12) billion years from point A to earth, blue light has become red light.

Since the majority of the radiation from ordinary stars and galaxies is centered around the spectrum of light, it is obvious that most of this radiation becomes microwave radiation by the time it reaches earth after traveling more than twelve (12) billion years. In thirty (30) billion years, blue light becomes 1123 nm in wavelength as readily seen by Figure 3.

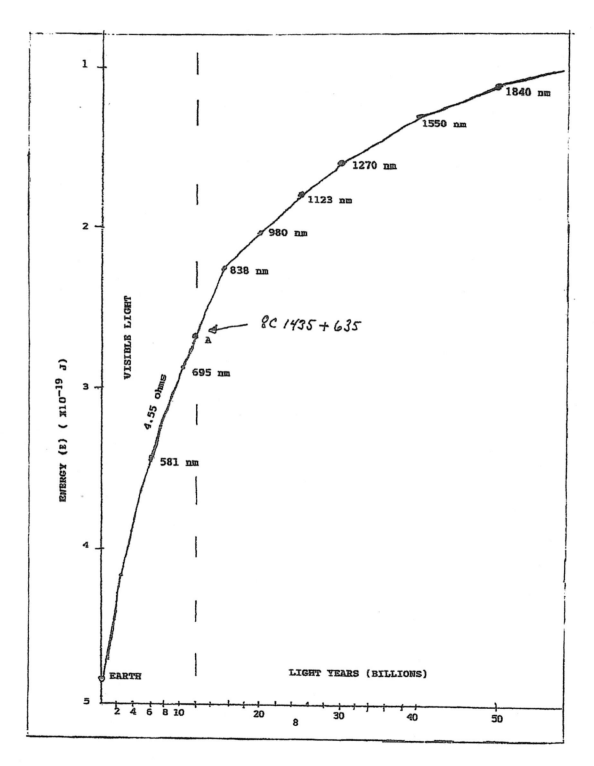

Figure 3. Plots the 410.2 nm wavelength with energy levels from one to five x 10-19 J.

On the Distribution of the Elements in the Universe

Physicists have long thought that the universe was young with an age less than 20 billion years. The predominance of hydrogen as the major constituent of all the elements was a powerful influence in their thinking. If the universe extended infinitely into the past, then the majority of the hydrogen should have condensed into stars.

In turn, the stars convert this basic element into all the other elements. Helium, oxygen, silicon, carbon and finally iron are produced in the fusion process. The amount of each element produced and then subsequently fused into other elements depends on the size and life history of the star.

Elements heavier than iron are created when the star dies in a sudden cataclysmic explosion. Physicists, therefore think that an infinitely aged universe would have little hydrogen and a much larger percentage of helium, oxygen, carbon, silicon and iron.

At the time this thinking was formulated and widely adopted by the scientific community, there was little reason to reject the idea. It seemed a reasonable hypothesis and hard to argue against. The hypothesis has now congealed and hardened among virtually all scientists.

In the early 1980's, I happened to see a photograph of two stellar objects called 3C273 and M87. I was also interested in black holes. Since I had never fully let go of a childhood belief that the universe is infinitely aged, I wondered if these objects might not explain the distribution of the elements in such a universe.

The hypothesis can be made as follows:

Just as the stars accrete hydrogen and generate the other heavier elements, then black holes accrete all elements where they are reduced to their basic constituents of electrons, protons, neutrons and perhaps even to quarks themselves by the immense gravitational and magnetic forces therein. The strong force itself is overwhelmed by the powerful gravitational and magnetic forces near a black hole.

In some cases, perhaps in 3C273 or some other stellar like objects such as quasars, these basic constituents are ejected out the polar regions of these objects in radio jets.

Article 9. THE PERIODIC TABLE
By Carrel W. Uptergrove

The periodic table is made of protons, neutrons and electrons in various combinations of each. Each combination has a name such as hydrogen (one proton and one electron), helium (two protons, two electrons and usually two neutrons).

There are more than one hundred elements. All of these elements except hydrogen and helium are created in stars in a process called fusion. We call them the ashes of stars. We are made from the ashes of stars. Usually every proton also has a neutron in the nucleus.

In space the elements are attracted to each other by gravity. When gravity becomes intense enough the star ignites in a fusion process which creates all the elements in the periodic table.

Black holes are very common in the universe. When stars are close enough to be within the magnetic lines of force of the black hole they are eventually drawn into the black hole. When there are no more stars to devour the black hole no longer feels the drag of the stars and its spin now increases.

The forces within the black hole are strong enough to reduce the elements created by the stars to be reduced back to hydrogen and helium. This hydrogen and helium can now be seen by the powerful telescopes we have today being ejected out of the polar regions of the black hole. Eventually the black hole is completely disappeared.

Now the process begins all over again. The tug of conflict between gravity and electromagnetism.

Printed in the United States
By Bookmasters